Teach Me the Earth's Atmosphere

AUTHOR AND ILLUSTRATOR TAMIKA K. FORDHAM

Teach Me the Earth's Atmosphere

AUTHOR AND ILLUSTRATOR TAMIKA K. FORDHAM

Copyright © 2012 Author Name

All rights reserved.

ISBN-13: 978-1548723606

ISBN-10:1548723606

Do you know that Earth has an **atmosphere**? Yes, it is true! The Earth's atmosphere is a mixture of gases. It includes water vapor, nitrogen, carbon dioxide, oxygen and more. In the atmosphere closest to us, there is more nitrogen, than oxygen. Nitrogen makes up 78% and oxygen make up 21%, equaling 99% of the atmosphere. The other 1% of gases is there in smaller amounts.

Teach Me the Earth's Atmosphere

The gases are thickest near the Earth's surface, where we live. This helps all plants grow and animals like you and me to breath in oxygen. In this layer, the water cycles over and over. Clouds are formed and we experience different types of storms, feel the wind blow up against our skin, and breathe in fresh air. The atmosphere is an important part of what makes life livable on Earth.

Teach Me the Earth's Atmosphere

Earth's atmosphere is made up of five layers. **Nitrogen** and **oxygen** is found in all five layers. The layer closest to Earth's surface is called the troposphere. Above the troposphere is the stratosphere. The third layer is called the mesosphere. Next to the last layer is the thermosphere. The layer farthest from Earth is called the Exosphere.

Teach Me the Earth's Atmosphere

In the layer closest to Earth's surface where we live, is called the **troposphere**. This is the layer where all of Earth's weather takes place, airplanes fly, and meteorologists use weather instruments to predict the weather every day! The atmosphere traps heat, making Earth temperatures comfortable for us to live and plants to grow.

Teach Me the Earth's Atmosphere

The **Stratosphere layer** is above the troposphere. It is the coldest an driest layer in Earth's atmosphere. The ozone layer is found here. The ozone layer protects Earth from the ultraviolet radiation from the Sun. Meteorologist sends out weather balloons between the troposphere and this layer, daily to predict the weather.

Teach Me the Earth's Atmosphere

In the **mesosphere** layer, you can observe meteorites burning up, exploding, or breaking apart. Sometimes we call them shooting stars. This layer of the atmosphere protects the Earth's surface from craters, like the moon's surface. Do you know why the moon has so many craters? If you said the moon doesn't have an atmosphere, then you are correct!

Teach Me the Earth's Atmosphere

The layer above the mesosphere is called the **thermosphere**. The air is very hot and thin here. In this layer, satellites orbit Earth's atmosphere. Satellites are used so that we can use technology devices; such as radios, cellular phones, weather radars, and televisions.

Teach Me the Earth's Atmosphere

The fifth layer of the Earth's atmosphere is the **exosphere** where space begins. This layer includes the Sun, our moon, and planets in our solar system, the asteroid belt, and outer space.

Teach Me the Earth's Atmosphere

The atmosphere blocks some of the Sun's dangerous rays from reaching Earth. The Earth's atmosphere is a thin layer of gases held in place by **gravity**. Do you know where Earth's atmosphere receive it's gravity to hold it in place? If you said the Sun, then you are correct!

ESSENTIAL QUESTIONS:

1. If Earth's atmosphere had another layer, what would you want it to be? What would you name it? What type of gases would located there? Short answer.

2. Illustrate a model to represent the composition of Earth's atmosphere.

Teach Me the Earth's Atmosphere

ESSENTIAL QUESTION:

1. Research the amount of gases in the Mesosphere that contribute to burning away meteors. How could knowing this help people on Earth?

ABOUT THE AUTHOR

Tamika K. Fordham is an author and educator. Presently a science teacher, she reads science books to her students to assist in, increasing their knowledge of what's being learned. As she teaches her students, parents, and conduct workshops, audiences rave at the excitement and enthusiasm she puts into her presentations that brings literacy in science alive. Through stories, it allows her students to use their imaginations, as she brings those stories to life. Much of her inspiration comes from teaching using anchor charts, childhood drawing competitions, and her imagination.

Tamika K. Fordham was born in Charleston, South Carolina. She graduated from South Carolina State University with her Bachelors and Masters of Arts in Teaching. She received her Educational Specialist Degree in Leadership from Liberty University. She furthered her certification, when she received her National Board Certification. She lives happily with her husband and son in South Carolina.

ACKNOWLEDGMENTS

This book is dedicated to my loving family who has always believed in me. I would like to thank you for pushing me to follow my dreams. This book was inspired by my teaching style, which is full of visuals, background knowledge and content, informational text, fascinating lessons, and knowing the interest of my students. To my family, students, and colleagues; I thank you for inspiring me and giving me the tools needed to achieve my goals. My science books are meant to educate children ages 4-12 years old. Enjoy reading about science that is a part of your everyday life.

Thank you Eartha, Richard, Kendrick, Michelle, Marion, and Shirley!

www.ingramcontent.com/pod-product-compliance
Lightning Source LLC
Chambersburg PA
CBHW041122180526
45172CB00001B/373